iPhone XS Manual for Techies

Master the Newest iPhone in an Hour or Less

© Copyright 2018 - All rights reserved.

The content contained within this book may not be reproduced, duplicated or transmitted without direct written permission from the author or the publisher.

Under no circumstances will any blame or legal responsibility be held against the publisher, or author, for any damages, reparation, or monetary loss due to the information contained within this book. Either directly or indirectly.

Legal Notice:

This book is copyright protected. This book is only for personal use. You cannot amend, distribute, sell, use, quote or paraphrase any part, or the content within this book, without the consent of the author or publisher.

Disclaimer Notice:

Please note the information contained within this document is for educational and entertainment purposes only. All effort has been executed to present accurate, up to date, and reliable, complete information. No warranties of any kind are declared or implied. Readers acknowledge that the author is not engaging in the rendering of legal, financial, medical or professional advice. The content within this book has been derived from various sources. Please consult a licensed professional before attempting any techniques outlined in this book.

By reading this document, the reader agrees that under no circumstances is the author responsible for any losses, direct or indirect, which are incurred as a result of the use of information contained within this document, including, but not limited to, — errors, omissions, or inaccuracies.

Table of Contents

Chapter One: Introduction 1

Chapter Two: iPhone XS description 5

   The Specifications of the Phone are Defined Below: 8

      Colors: 8

      Capacity: 8

      Dimensions: 9

      Weight: 9

      Display: 9

      Water and Dust Resistant: 10

      Chip: 10

      Face ID 10

      Location 10

      Power and Battery 10

      Sensors 12

      Operating System 12

      Accessibility 13

      Environmental Requirements 13

      Comparison 17

Chapter Three: iPhone Basics 21

   In the Box 21

   Setting up Your Iphone XS 21

    Gesture Controls 21

    The Basic Gestures: 22

Setting up Apple Pay 23

Configuration of iCloud 24

Chapter Four: Synching Data 25

You're Setting up Your First iOS Device 25

Backup Old iPhone & Restore to iPhone XS using iCloud ... 26

Setting up iTunes 28

Sync Content using Wi-Fi 29

Chapter Five: Phone Call Basics 31

Face ID 31

Setting up an E-Sim 32

Cellular and Wireless 34

Location 35

Video Calling 35

Audio Calling 35

SIM Card 36

Chapter Six: Texting, Notes, and Utilities 37

Text Messages 37

iMessage 37

Animoji 39

Digital Touch Feature: 39

GIFS 40

Apple Music.. 40

Notes 40

Utilities 43

Chapter Seven: Phone Customization.. 45

Chapter Eight: Multimedia.. 49

Camera... 50

Display... 52

Video.. 53

True Depth Camera 54

Audio Playback 55

Video Playback... 55

Chapter Nine: Using the Internet 57

Wi-Fi Hotspot... 57

Safari 58

Mail 60

Mail Attachment Support.. 60

To View Mail Settings.. 62

Chapter Ten: Phone Settings 65

Turn on Battery Percentage... 65

How to Reset to Factory Settings 65

How to Set Screen Lock Time 66

Hide Unwanted Apps 66

Change Flashlight Intensity 67

Auto Delete Unused Apps 67

Emoji Keyboard Replacement 67

Prioritize Certain Downloads 68

How to Set an iPhone Email Signature 68

Change the Wallpaper on Your Device 68

Organizing Apps 68

Creating a Folder 69

Chapter Eleven: Apps 71

Chapter Twelve: Tips, Tricks, and Shortcuts 77

Chapter Thirteen: Conclusion 83

Chapter One
Introduction

Endeavoring the world of electronics, in particular, that of mobile phones and computer servers, every one of us know who the market leader is and who has the most well-established repute in the market. Yes, we are talking about Apple Inc, formerly known as Apple Computer Inc. However, this success did not come to Apple easy; it took a great deal of hard work to lay the rock-hard foundation of what is today the world's biggest and most trusted brand. It's been 42 years since Apple was founded.

From the garage of Steve jobs to having a retail outlet in every city of the world, Apple has come a long way. Let us go back to April 1, 1986. The three owners, namely Steve jobs, Ron Wayne, and Steve Wozniak united over the dream of shrinking room-sized computers to something more compatible and more practical to use. However, it was not long before Ron Wayne sold his shares and decided to leave Apple Inc. It's ironic how Jobs and Wozniak started their company by launching Apple 1, a circuit keyboard without a case or a keyboard, and now it's launching the state of the art iPhone X Max.

The first revolutionary smartphone by Apple was announced in June 2007. From that day onwards the 'iPhone' has forever changed our perspective of cell phones. Recently, Apple launched one of its most successful iPhones, called the 'iPhone XS.' Now this phone is the successor to the iPhone X and has a lot of new features, as Apple always excites us.

To make it easy for the audience to assess the new phone, we will be making a short and to the point manual for easy understanding for which you don't have to be tech-savvy. Now who likes going on the internet to research a phone and waste a lot of time and effort only to find that there is a lot of junk information? No one does! The internet is a deep place where you don't know whether you'll get the right information or not. So, in order to save you the effort and make it fun as well in learning what the new iPhone has to offer, we welcome you to check out our 'iPhone XS Manual.' Next you can find out the differences between the old and new phones to see how willing you are to upgrade. The manual is available in 10 languages so is easily accessible to everyone. For people to judge the new technology easily, we have given all of the major specifications of the phone and how they are going to work better than the previous one and help you in your daily life. In the end, it's all about how technology is going to make our lives more convenient and that is briefly explained inside as well to help make up your mind. Lastly, who wants to give up watching an appealing manual with easily explained features? The answer is, no one. With the iPhone XS, you are getting the best phone designed with the best features Apple has created so far. Every year the

Introduction

iPhone gets better and better, and now it takes another leap. Learn as you go through this manual.

In this manual, you will learn ways to maximize the use of the iPhone XS. It describes how to quickly get started with your device, and once you complete reading the manual, you will know everything that you need to know about the iPhone XS. From the basic settings to all the hidden features of the iPhone XS, every single function will be at your fingertips. This manual includes tips and tricks that will be helpful towards making your device more useful in your everyday life. This extensive user manual includes all the instructions that will enable you to use the iPhone XS effectively, and since iPhones don't come with a printed user guide, this manual is going to be the only thing that you need. It's easy to understand and is user-friendly, for the techno-savvy to a first time iOS user; this manual can be used by anyone across the board.

Chapter Two
iPhone XS description

Apple is one of the top brands when it comes to smartphones. It has proven each and every time to set the bar high for other companies in the market, and yet again this year they came up with a superb upgrade. Despite being an upgraded version of the iPhone X that was launched last year, the iPhone XS has some significant advancements in the phone.

The iPhone XS is the future of smartphones. After the launch of the iPhone XS, it can be clearly seen that Apple has made some significant advancement across the board. It comes with a sleek design and perfectly crafted corners. Despite being an upgrade of the iPhone X, there are many remarkable features that did not fail to amuse us. The iPhone XS is almost the same as the iPhone X; if you put them side by side, it's very hard to tell the difference between them, except for a few key things, like the new color gold. It's more of a saturated gold which makes it distinct from the previous models. It's a very polarizing color, especially the stainless-steel sides which are really gold. They are the most striking part of the iPhone.

Apple has always been known for its glamorous iPhones and how they always have incremental upgrades that they launch each year. No matter what, Apple never fails to outshine all other brands. The sleek and stylish design has pleased everyone around the globe. It not only comes with super retina HD display, combined with a high dynamic range, it has the largest display ever for an iPhone. Multi-touchscreen incorporates multi-touch technology. Multi-touch allows users to use more than one finger simultaneously to perform functions on the screen. Some of the most talked about features of the iPhone are due to the multi-touch proficiency, such as tapping the screen twice to zoom in or "pinching" and dragging your fingers to zoom out

The iPhone XS has two more visible antennas than previous models. They are located on the stainless-steel bands around the outside. One at the top right above the camera module and one down at the bottom, kind of cutting into the speaker grill. And now the speaker grill at the bottom is asymmetrical because of the antenna band.

Other than this, all other hardware features are the same, such as same button placement, same curved corner, same weight, glass sandwich phone, same mute switch, same notch, and same fingerprint magnet glass. Another small improvement is better speakers. Louder and clearer stereo is a noticeable improvement. With the iPhone XS it is harder to take accidental screenshots. So previously, if you woke up and unlocked your iPhone with the power button and the volume button at the same time, it often took a screenshot, but now it will not take a screenshot of the home screen.

Technically, the iPhone XS is ip68 water resistant where the iPhone 10 was ip67. It's a huge difference as ip67 means you can take your phone one meter under water for a straight thirty minutes, whereas ip68 water resistance meant you could take your iPhone XS 1.5 meters for thirty straight minutes. It's 15-30% faster in performing several actions like opening apps, launching the cameras, and opening the keyboard when compared to all previous models. As a matter of fact, this upgrade actually makes it the king of responsiveness as far as smartphones go. iOS 12 is so buttery smooth that it is back on the throne! So, people with an older phone, when they upgrade to iOS 12, they will not experience a lag and will noticeably see that their phone is working faster.

As far as the bugs are concerned, none of them are crashing the phone or closing the apps or even lagging them. There's a slight glitch in the new resolution, overlapping graphics and some cutting of graphics due to the notch, but they are small errors that will be fixed quickly.

The iPhone XS has an A12 bionic chip, which has 7-nanometer technology with an 8-core devoted machine-learning engine to evaluate neural network data to presume whether the processes should run through the neural engine or not. The A12 bionic processor uses less energy and can perform 5 trillion operations per second, giving it the ability to open apps 30 times faster than previous iPhones. Being one of the first chips on the market to possess this technology, this chip has some insane advantages. The first and foremost feature that is important to us is the performance and speed of the device. With the A12 bionic chip, we will get a 25% performance boost, which means the phone will be running 25% faster, and we will also get a 30% graphic boost. Moreover, the battery life is also improved by this technology. We will also get better and more efficient encoding for videos, slow-motion videos, 4K, and everything along those lines, which means that everything will be sharper, crisper, and will be running smoother.

The Specifications of the Phone are Defined Below:

Colors:

- gold,
- space gray
- silver

Capacity:

- 64GB
- 256GB
- 512GB

iPhone XS description

Dimensions:

- Height 6.20 inches (157.5mm)
- Width 3.05(77.4mm)
- Depth 0.30 inches (7.7mm)

Weight:

- 7.34 ounces (208 grams)

Display:

- Super retina display (HD)
- 6.5 inch (when measured diagonally) all screen OLED
- Multi touch
- HDR display
- 2688-by-1242-pixel resolution at 458 ppi
- 1,000,000:1 contrast ratio (typical)
- True Tone display
- Wide color display (P3)
- 3D Touch
- 625 cd/m2 max brightness (typical)
- Fingerprint-resistant oleo phobic coating
- Support for display of multiple languages and characters simultaneously

Water and Dust Resistant:

- Rated IP68 under IEC standard 60529

Chip:

- A12 Bionic chip
- Next-generation Neural Engine
- Wide-angle: $f/1.8$ aperture
- Telephoto: $f/2.4$ aperture
- Extended dynamic range for video at 30 fps

Face ID

- Enabled by TrueDepth camera for facial recognition

Location

- Assisted GPS, GLONASS, Galileo, and QZSS
- Digital compass
- Wi-Fi
- Cellular
- iBeacon microlocation

Power and Battery

- Lasts up to 30 minutes longer than iPhone X
- Talk time (wireless):

iPhone XS description

- Up to 20 hours
- Internet use:
- Up to 12 hours
- Video playback (wireless):
- Up to 14 hours
- Audio playback (wireless):
- Up to 60 hours
- Fast-charge capable:
- Up to 50% charge in 30 minutes
- Lasts up to 1.5 hours longer than iPhone X
- Talk time (wireless):
- Up to 25 hours
- Internet use:
- Up to 13 hours
- Video playback (wireless):
- Up to 15 hours
- Audio playback (wireless):
- Up to 65 hours
- Fast-charge capable:
- Up to 50% charge 30 minutes8

- Both models:
- Built-in rechargeable lithium-ion battery
- Wireless charging (works with Qi chargers)
- Charging via USB to computer system or power adapter

Sensors

- Face ID
- Barometer
- Three-axis gyro
- Accelerometer
- Proximity sensor
- Ambient light sensor

Operating System

- iOS 12
- iOS is the world's most personal and secure mobile operating system, packed with powerful features that help you get the most out of everyday use.
- See what's new in iOS 12

Accessibility

- Accessibility features help people with disabilities get the most out of their new iPhone XS. With built-in support for vision, hearing, physical and motor skills, and learning and literacy, you can fully enjoy the world's most personal device

Environmental Requirements

- Operating ambient temperature:
- 32° to 95° F (0° to 35° C)
- Nonoperating temperature:
- −4° to 113° F (−20° to 45° C)
- Relative humidity:
- 5% to 95% noncondensing
- Operating altitude:
- tested up to 10,000 feet (3000 m)
- Languages
- Language support
- English (Australia, UK, U.S.), Chinese (Simplified, Traditional, Traditional Hong Kong), French (Canada, France), German, Italian, Japanese, Korean, Spanish (Latin America, Mexico, Spain), Arabic, Catalan, Croatian, Czech, Danish, Dutch, Finnish, Greek,

Hebrew, Hindi, Hungarian, Indonesian, Malay, Norwegian, Polish, Portuguese (Brazil, Portugal), Romanian, Russian, Slovak, Swedish, Thai, Turkish, Ukrainian, Vietnamese

- QuickType keyboard support

- English (Australia, Canada, India, Singapore, UK, U.S.), Chinese - Simplified (Handwriting, Pinyin, Stroke), Chinese - Traditional (Cangjie, Handwriting, Pinyin, Stroke, Sucheng, Zhuyin), French (Belgium, Canada, France, Switzerland), German (Austria, Germany, Switzerland), Italian, Japanese (Kana, Romaji), Korean, Spanish (Latin America, Mexico, Spain), Arabic (Modern Standard, Najdi), Armenian, Azerbaijani, Belarusian, Bengali, Bulgarian, Catalan, Cherokee, Croatian, Czech, Danish, Dutch, Emoji, Estonian, Filipino, Finnish, Flemish, Georgian, Greek, Gujarati, Hawaiian, Hebrew, Hindi (Devanagari, Transliteration), Hinglish, Hungarian, Icelandic, Indonesian, Irish, Kannada, Latvian, Lithuanian, Macedonian, Malay, Malayalam, Maori, Marathi, Norwegian, Odia, Persian, Polish, Portuguese (Brazil, Portugal), Punjabi, Romanian, Russian, Serbian (Cyrillic, Latin), Slovak, Slovenian, Swahili, Swedish, Tamil (Script, Transliteration), Telugu, Thai, Tibetan, Turkish, Ukrainian, Urdu, Vietnamese, Welsh

- Quick Type keyboard support with predictive input

iPhone XS description

- English (Australia, Canada, India, Singapore, UK, U.S.), Chinese (Simplified, Traditional), French (Belgium, Canada, France, Switzerland), German (Austria, Germany, Switzerland), Italian, Japanese, Korean, Russian, Spanish (Latin America, Mexico, Spain), Portuguese (Brazil, Portugal), Thai, Turkish

- Siri languages

- English (Australia, Canada, India, Ireland, New Zealand, Singapore, South Africa, UK, U.S.), Spanish (Chile, Mexico, Spain, U.S.), French (Belgium, Canada, France, Switzerland), German (Austria, Germany, Switzerland), Italian (Italy, Switzerland), Japanese, Korean, Mandarin (Mainland China, Taiwan), Cantonese (Mainland China, Hong Kong), Arabic (Saudi Arabia, United Arab Emirates), Danish (Denmark), Dutch (Belgium, Netherlands), Finnish (Finland), Hebrew (Israel), Malay (Malaysia), Norwegian (Norway), Portuguese (Brazil), Russian (Russia), Swedish (Sweden), Thai (Thailand), Turkish (Turkey)

- Dictation languages

- English (Australia, Canada, India, Indonesia, Ireland, Malaysia, New Zealand, Philippines, Saudi Arabia, Singapore, South Africa, United Arab Emirates, UK, U.S.), Spanish (Argentina, Chile, Colombia, Costa Rica, Dominican Republic, Ecuador, El Salvador,

Guatemala, Honduras, Mexico, Panama, Paraguay, Peru, Spain, Uruguay, U.S.), French (Belgium, Canada, France, Luxembourg, Switzerland), German (Austria, Germany, Luxembourg, Switzerland), Italian (Italy, Switzerland), Japanese, Korean, Mandarin (Mainland China, Taiwan), Cantonese (Mainland China, Hong Kong, Macao), Arabic (Kuwait, Qatar, Saudi Arabia, United Arab Emirates), Catalan, Croatian, Czech, Danish, Dutch (Belgium, Netherlands), Finnish, Greek, Hebrew, Hindi (India), Hungarian, Indonesian, Malaysian, Norwegian, Polish, Portuguese (Brazil, Portugal), Romanian, Russian, Shanghainese (Mainland China), Slovakian, Swedish, Thai, Turkish, Ukrainian, Vietnamese

- Definition dictionary support

- English, Chinese (Simplified, Traditional), Danish, Dutch, French, German, Hebrew, Hindi, Italian, Japanese, Korean, Norwegian, Portuguese, Russian, Spanish, Swedish, Thai, Turkish

- Thesaurus

- English (UK, U.S.)

- Bilingual dictionary support in English

- Arabic, Chinese (Simplified, Traditional), Dutch, French, German, Hindi, Italian, Japanese, Korean, Portuguese, Russian, Spanish

- Spell check
- English, French, German, Italian, Spanish, Danish, Dutch, Finnish, Korean, Norwegian, Polish, Portuguese, Russian, Swedish, Turkish

Taken from https://www.apple.com/iphone-xs/specs/

As we know that the iPhone XS is a flagship phone from Apple and is an upgrade of the iPhone X, let's see why the iPhone XS Max is the best amongst the X series.

Comparison

SPECS	iPhone XS	iPhone XS Max	iPhone XR
Colors	Silver, black, gold	Silver, black, gold	Blue, white, red, black, yellow, coral, red
Display	5.8" Super retina display	6.5" Super retina display	6.1" Liquid Retina HD display
Dual	12MP wide-angle and telephoto cameras	12MP wide-angle and telephoto cameras	12MP wide-angle and telephoto cameras
TrueDepth camera	7MPTrueDepth camera	7MPTrueDepth camera	7MPTrueDepth camera
Face ID	Yes	Yes	Yes
Chip	A12 Bionic chip with next-generation Neural Engine	A12 Bionic chip with next-generation Neural Engine	A12 Bionic chip with next-generation Neural Engine
Water resistant	2mfor up to 30 minutes	2mfor up to 30 minutes	1mfor up to 30 minutes
Charging	Wireless charging	Wireless charging	Wireless charging
Capacity	64,256,512GB	64,256,512GB	64,128,256GB

iPhone XS Manual for Techies

Display	-HDR display -True Tone display -Wide color display -3D Touch -625 cd/m2 max brightness	-HDR display -True Tone display -Wide color display -3D Touch -625 cd/m2 max brightness	-True Tone display -Wide color display -625 cd/m2 max brightness (typical)
Wireless & cellular	-Bluetooth 5.0 -GPS, GLONASS, Galileo, and QZSS NFC with reader mode -Express Cards with power reserve	-Bluetooth 5.0 -GPS, GLONASS, Galileo, and QZSS NFC with reader mode -Express Cards with power reserve	-Bluetooth 5.0 -GPS, GLONASS, Galileo, and QZSS NFC with reader mode -Express Cards with power reserve
Video Calling	FaceTime video	FaceTime video	FaceTime video
Audio Playback	Wider stereo playback User-configurable maximum volume limit	Wider stereo playback User-configurable maximum volume limit	Wider stereo playback User-configurable maximum volume limit
Power and Battery	-Lasts up to 30 minutes longer than the iPhone X -Built-in rechargeable	-Lasts up to 1.5 hours longer than the iPhone X -Built-in rechargeable lithium-ion battery -Charging via USB to computer system or power adapter	-Lasts up to 1.5 hours longer than the iPhone 8 Plus -Built-in rechargeable lithium-ion battery
Internet use	Up to 12 hours	Up to 13 hours	Up to 15 hours
Video playback	Up to 14 hours	Up to 15 hours	Up to 16 hours
SPECS	iPhone XS	iPhone XS Max	iPhone XR

iPhone XS description

Video formats	-HEVC, H.264, MPEG-4 Part 2, and Motion JPEG -High Dynamic Range with Dolby Vision and HDR10 content	-HEVC, H.264, MPEG-4 Part 2, and Motion JPEG -High Dynamic Range with Dolby Vision and HDR10 content	-HEVC, H.264, MPEG-4 Part 2, and Motion JPEG -High Dynamic Range with Dolby Vision and HDR10 content
Video mirroring and video out support	Up to 1080p through Lightning Digital AV Adapter and Lightning to VGA Adapter	Up to 1080p through Lightning Digital AV Adapter and Lightning to VGA Adapter	Up to 1080p through Lightning Digital AV Adapter and Lightning to VGA Adapter (adapters sold separately)
Audio playback	Up to 60 hours	Up to 65 hours	Up to 65 hours
Talk time	Up to 20 hours	Up to 25 hours	Up to 25 hours
Fast-charge capable	Up to 50% charge 30 minutes	Up to 50% charge 30 minutes	Up to 50% charge 30 minutes
Headphones	EarPods with Lightning Connector	EarPods with Lightning Connector	EarPods with Lightning Connector
Sensors	-Three-axis gyro -Accelerometer -Proximity sensor -Ambient light -Barometer	-Three-axis gyro -Accelerometer -Proximity sensor -Ambient light -Barometer	-Three-axis gyro -Accelerometer -Proximity sensor -Ambient light -Barometer
SIM Card	-Dual SIM (nano-SIM and eSIM) -Not compatible with existing micro-SIM cards	-Dual SIM (nano-SIM and eSIM) -Not compatible with existing micro-SIM cards	-Dual SIM (nano-SIM and eSIM) -Not compatible with existing micro-SIM cards

Chapter Three
iPhone Basics

In the Box

Getting started with your new iPhone XS. Once you get an iPhone XS, while unboxing you will find the following:

- iPhone with iOS 12
- Ear Pods with Lightning Connector
- Lightning to USB Cable
- USB Power Adapter
- Documentation

Setting up Your Iphone XS

Gesture Controls

Learning gestures is one the most basic things to do once you get your iPhone XS. Gesture controls and navigating around the iPhone without the home button can be quiet challenging in the beginning, so here are the gestures for you to be a pro.

The Basic Gestures:

- Swiping up from the bottom gets you to the home screen, so if you're in an application and you want to get out then swipe up from the bottom and you will return to the home screen.

- Bring down the control center by swiping down from the top right. If you do it from the top middle it will view the notification center.

- If you want to see which applications you have running in the background, then just swipe up and hold, it will bring up the background running applications, and if you want to kill the applications running in the background just swipe up. Now you don't have to wait for the red sign to pop up in order to kill the apps and it is far more convenient.

- If you want to evoke Siri, hold the button on the right side and this command will activate Siri.

- If you want to turn the phone off, press the home button up, the home button down, and then hold the power button.

- To use Apple Pay, double click the side button to view the credit card information that you have entered. Then if you have enabled Face ID for Apple Pay then use Face ID to authenticate.

- In order to use accessibility shortcut, triple click the side button.

- To make an emergency call, press and hold the side button and volume button at the same time, sliders will appear, then drag emergency SOS to make the call.

- To force restart, press and release the volume up button then press and release the volume down button and then press and hold the side button until the Apple logo appears

Setting up Apple Pay

- Tap to wallet
- Add card/scan card (e.g. card coupons, membership cards, coupons, reward cards, passes)
- Scan your debit/credit card so you can securely pay while online shopping

It will create ease for you instantly, no hassle of carrying cards or other passes with you. Scanning makes it even more convenient since you don't have to enter the information manually, best for those who are in a rush or on the go.

Configuration of iCloud

- Go to settings

- Sign in to your iCloud account/sign up

- Once signed in tap on the iCloud and customize the settings

- Sync the data that you want to backup (photos, mail, contacts, safari news, stocks, home, calendar, reminders, notes, messages, health, wallet, game center, Siri

- Customize other settings as well, including find my iPhone, send my last location in case of theft, or if you have misplaced your iPhone, these features come in handy

If you have a lot of apps, music, and other files that you really want to move from your old phone to your new device, you can restore from the back up either from iTunes or iCloud.

Chapter Four
Synching Data

You can transfer your data from your old iPhone to the new iPhone XS, which includes all the contacts, text messages, applications, music, photos, memos, basically everything. When you power on your new iPhone and select your language and country you'll be prompted to either

- Restore your iPhone from a backup

- Set up your device as a new iPhone

- Move data from an android device

You're Setting up Your First iOS Device

- Switch on the iPhone XS (press and hold the power button)

- Select the country where you're residing

- Set up iPhone XS manually

- Choose Wi-Fi network/carrier internet connection

- Activate the iPhone

- Set up Face ID by following the instructions that appear on the screen, or otherwise skip this step and create a passcode for the device

- Select "set up as new iPhone" option or move data from Android

- Create a free Apple ID

- Select the suitable option for automatic updates or manual (Recommended settings would be the manual option since it conserves battery)

- Set up iMessage and Facetime

- Enable location services

- Enable Siri

- Set up screen time, app analytics (if you want to share data with app developers)

- Select true tone display to be on or off. You can see the difference by clicking the "with and without" button and then choose accordingly

Backup Old iPhone & Restore to iPhone XS using iCloud

- Choose the language of your iPhone XS

- Select the country where you're residing

- If you have a previous iPhone or iPad running on iOS 11 or later, bring it nearby to sign in automatically or otherwise you can set it up manually

Synching Data

- Once you bring the old device near, lock and unlock it. A prompt message will show appear prompting you to "set up new iPhone." Press continue to start the setup process

- Once the two phones connect, put the old iPhone in the frame of the camera of iPhone XS. Once it's successfully done, a message will appear on the iPhone XS prompting you to "enter passcode of your other iPhone"

- Enter the passcode, and iPhone XS will start the setup process. Meanwhile, keep the old iPhone near and make sure it doesn't lock. The iPhone XS will then attempt to activate the phone. It will even automatically pull in the Wi-Fi!

- A flash screen will appear of "data and privacy." Read it carefully and then press continue, but if there are any concerns regarding your data and privacy, you can always click on the "learn more" option

- The next step would be setting up Face ID, in order to unlock the iPhone XS. Follow the instructions that pop up on the screen to set up Face ID. (If you are in a rush you can set it up later and skip to the home screen)

- After reading the terms and conditions to use the iPhone XS, press the continue button

- Now customize the settings which you want to back up

- The next page will be "Keeping your phone up to date." By enabling this feature your iPhone XS will automatically update when an update is available, or otherwise you can select the install updates manually option (Recommended settings would be the manual option since it conserves battery)

- Apple Pay, screen time, app analytics (if you want to share data with app developers) are the setup options that you are required to set up

- Select true tone display to be on or off. You can see the difference by clicking the "with and without" button and then choose accordingly

- Enter the password for your iCloud or iTunes to back up the data

- The restore process should not take too long, but it will mainly depend upon your network speed and will also depend on how much data you are transferring

Setting up iTunes

- Connect your iPhone to a computer using the lightning connector USB cable

- Open iTunes if it doesn't open automatically

- iTunes will greet you and will get you started
- Select iPhone XS in the list of devices in the left panel
- Click the iPhone summary tab which will tell you about your device and will give you information about your backups
- Select restore backup
- If you have backups of multiple devices, you will be able to tell which one is the most recent by the time and date stamp
- After you restore the backup, sync your phone to your computer
- Then eject the device

Note: Keep in mind that the backup needs to be on the same version or lower version of the current phone

Sync Content using Wi-Fi

1. Connect your iPhone to a computer using the lightning connector USB cable
2. Open iTunes if it doesn't open automatically
3. iTunes will greet you and will get you started
4. Select iPhone XS in the list of devices in the left panel
5. Select "Sync with iPhone XS over Wi-Fi."
6. Click Apply

When your computer and iPhone are connected over the Wi-Fi network, your device will appear in iTunes. The benefit of syncing your data over Wi-Fi, is that the device automatically syncs data whenever the device is connected to the Wi-Fi and iTunes is open on the computer at the same time.

Chapter Five
Phone Call Basics

Face ID

- Face ID is so incredible with the hottest feature of group calling friends up to thirty-two people, making it better than ever. Another exciting feature of Face ID is its password autofill option. So rather than entering passwords, enable Face ID for different apps to avoid the hassle of entering passwords each and every time

- Setup your face ID by taking a clear picture of your face, and also set up an alternative appearance (e.g. if you wear a lot of makeup or if you want your wife to get in your phone) Toggle all the options and see which ones are suitable whilst setting up. Face ID is Enabled by TrueDepth camera for facial recognition

- Face ID is faster in the iPhone XS. RGB cameras that are used in other smartphones are quicker than that to Apple, but the edge is that what Apple provides is more secure and reliable.

Setting up an E-Sim

- If you open the sim tray of your device, it will only have a single nano sim option available. The other sim is an electronic sim which is built into the device itself. In order to activate that sim you will need to either scan the QR code provided by your carrier or get the eSim through an app.

- Go to settings.

- Scroll and click on cellular and click on the option where you can add an additional plan. Click on "Add cellular plan."

- Scan the QR code that is provided by the carrier.

- Enter the confirmation code to activate the eSim.

- Once it is activated set up labels for both your sims.

- Select the default line. The default line would be the number which you want to use for making calls and sending messages

- "Use primary as your default line." If you select this option, your primary number will be used by default for voice, SMS, Data, iMessage, and FaceTime, and the secondary number will be available for voice calling and SMS.

- "Secondary as your default line." If you select this option, your secondary number will be used for voice,

SMS, Data, iMessage, and FaceTime whilst your primary number will be used for voice calling and SMS.

- "Use Secondary for cellular data only." If you select this option, you will keep your primary number for voice, SMS, iMessage, and FaceTime. And your second number for data consumption.

- If your carrier informs you to get the eSim through an app then the following applies:

- Download the required carrier app; it can be downloaded from the App Store

- Purchase the required eSim plan using the app, but before that you need to sign into the app before making the purchase.

As of this moment, only certain carriers provide eSim technology. To check whether your carrier supports the e-sim feature at this moment:

- Go to settings>>general>>about >>

- Check for the IMEI number. It will give a primary and secondary IMEI number. If the secondary IMEI number is enabled, it means that your carrier provides the facility of eSim

eSim supported Carriers:

- **Austria:** T-Mobile
- **Canada:** Bell

- **Croatia:** Hrvatski Telekom
- **Czech Republic:** T-Mobile
- **Germany:** Telekom, Vodafone
- **Hungary:** Magyar Telekom
- **India:** Reliace Jio, Airtel
- **Spain:** Vodafone Spain
- **United Kingdom:** EE
- **United States:** AT&T, T-Mobile USA, and Verizon Wireless

Cellular and Wireless

- Model A2097*
- FDD-LTE (Bands 1, 2, 3, 4, 5, 7, 8, 12, 13, 14, 17, 18, 19, 20, 25, 26, 28, 29, 30, 32, 66)
- TD-LTE (Bands 34, 38, 39, 40, 41, 46)
- UMTS/HSPA+/DC-HSDPA (850, 900, 1700/2100, 1900, 2100 MHz)
- GSM/EDGE (850, 900, 1800, 1900 MHz)
- Model A2101*
- FDD-LTE (Bands 1, 2, 3, 4, 5, 7, 8, 12, 13, 14, 17, 18, 19, 20, 25, 26, 28, 29, 30, 32, 66)

- TD-LTE (Bands 34, 38, 39, 40, 41, 46)
- UMTS/HSPA+/DC-HSDPA (850, 900, 1700/2100, 1900, 2100 MHz)
- GSM/EDGE (850, 900, 1800, 1900 MHz)
- Gigabit-class LTE with 4x4 MIMO and LAA4
- 802.11ac Wi-Fi with 2x2 MIMO
- Bluetooth 5.0 wireless technology
- NFC with reader mode
- Express Cards with power reserve

Location

- Assisted GPS, GLONASS, Galileo, and QZSS
- Digital compass
- Wi-Fi
- Cellular
- iBeacon micro location

Video Calling

- FaceTime video calling over Wi-Fi or cellular

Audio Calling

- Supports FaceTime audio only.

SIM Card

- Dual SIM (nano-SIM and eSIM)10
- The iPhone XS and iPhone XS Max are not compatible with existing micro-SIM cards.
- Rating for Hearing Aids
- M3, T4
- M3, T4

Chapter Six
Texting, Notes, and Utilities

Text Messages

- Launch the message application on your device
- Click on the compose button in the upper right corner
- Type the name of the recipient
- Type your message in the message field

iMessage

For years, iPhone users across the world have claimed that iMessage is the best feature in their Apple device. If you type in hello, and before sending it 3D touch it, a bunch of options will pop up

Here are different ways you can modify an iMessage:

- **Slam effect:** when the receiver gets the message the screen will quake from the slam effect
- **Loud**: make the message wiggle a bit
- **Gentle**: the message starts off in a very small text size and gets a little bit bigger
- **Invisible ink**: Initially the receiver will not be able to see the message, unless they wipe across the message to read it. After a few seconds it will revert to the

hidden view again. This effect not only works with messages but also works on images too

Other than the effects, iMessage also comes with another tab just next to the effects tab, the **"screen tab."** This basically changes what the entire messaging application does on the screen with sound and vibration effects that go along with them:

- **Echo**: The screen will show the message on the entire screen in an echo form and will pop on all over the place.

- **Spotlight**: It will throw light on the particular message and make the remaining screen black.

- **Balloons**: Sound of balloons filling up and balloons rubbing together, and once the receiver gets it, balloons will appear all over the screen.

- **Confetti**: Pops the confetti and it drops down along with the sound of popping and falling.

- **Heart**: The heart blows up on the entire screen.

- **Laser effect**: Laser light coming out from the sides of the message.

- **Fireworks**: Background is filled with fireworks and accompanied by their popping noise.

- **Shooting star**: A shooting star in the background once the message is delivered.

- **Celebration**: A sparkler effect in the background once the message is delivered.

The phone vibrates along with the effects. Different effects have different rhythms of vibration; it's more of a tapping sensation than a vibration

Animoji

- Animate the emoji
- Record ten seconds worth of audio
- Sends as a regular GIF to anyone who doesn't have an iPhone (it will have the audio as well)

Digital Touch Feature:

- Draw around on it, change colors
- Tap to make circles, you can change the colors as you go
- Tap with two fingers to make two little lips appear. Depending on where you put your fingers, the lips will align with them
- Hold with two fingers. You will see a little heartbeat, and once you let go it will send it or you can drag down and it breaks the heart
- 3D touch: a fireball appears. Once you force touch it you can drag it around

Note: Lips and fireballs do not change colors.

If you want to use a bunch of different types of digital touch features effects in one message, the only way you can do it is by taking a picture and animating over the top of it.

If you record a video you are going to have ten seconds of recording, you can tap to add circles and other digital touch features in the video, and after ten seconds it is going to finish.

GIFS

- A bunch of different GIFS are available.
- To find an appropriate GIF you can always search for it in the search bar.

Apple Music

This allows you to send a song to a friend. Tap the desired song you want to send.

Notes

In app features include adding tables while making notes to input data; you can rearrange the table by clicking and dragging the columns and the rows using the drag handles at the side. You can also add or remove the columns or rows by simply using the handles.

- If you long press the table icon several options will appear
  - Copy table
  - Share table
  - Convert to text
  - Delete table
- Texting Formatting redesign options are available, which makes the adjustment of text easier. Some options that are available once you long press the formatting icon are:
  - Monospaced
  - Bold
  - Underline
  - Bullets/numbering
  - Indentation options
- Checklists option is available to make a shopping list, to do list
- The plus icon will give you more options that you can use:
  - Scan documents option allows you to scan documents natively

- Take photo or video
- Photo library
- Add sketch

- Lines and grids can also be added within the note. There are different background options from which you can choose
- Sorting options for notes are available in its view settings. You can sort it by date edited, title, or date created
- Start new notes

Choose Default background for your notes. Password options are also available. You can create secure notes as well, by putting a lock on them. Launch the notes application and you want to create a new note which has sensitive information in it. After you have created your note, tap the share button (next to the done button) at the bottom of the screen. The lock note option will appear. Your device is going to ask you for Face ID if you have it enabled, or otherwise you can enter the password which you use to unlock your phone. After authenticating, a prompt will show up as "lock added." In the Notes app, it will only show the title of the note and not the content. If you want to view the note, tap on it and enter the passcode or authenticate with Face ID.

Utilities

- The iPhone XS is not just a smartphone, it has all the features that you could want in a device. It has the ability to manage your personal information with the help of applications like calendar, weather updates, stock management, address book, maps, and compass, etc.

- These apps can be used in our everyday lives to make it much easier. Call reminder is another feature that you can use to make your device remind you to make calls that you have missed. You can also choose a particular time of the day when you want these notifications. This feature is convenient for students, businessmen, and for anyone who tend to keep their phones away when they are busy.

Chapter Seven
Phone Customization

- Animoji uses the front facing camera to map your face to a 3D emoji. Animoji is inside of the iMessage app under applications, but you can use it in lots of places. It is pre-installed on the iPhone XS. You can also record a short ten-second video clip of your face making different gestures like blinking, talking, rotating your head, and animojis mimic the same expression with surprising accuracy. Customized animojis, can be set up with the help of the face recognition tool. Simply open the app drawer in the message app. Swipe right and select the Animoji (monkey) icon and swipe right until you see the New Memoji icon (+) which allows the user to customize as they like. After you are finished, tap the "Done" button at the top of the screen to save your Memoji and use them in your iMessages.

- Customize your device to unlock it without pressing the button. Tap to wake up the phone. In order to wake up the phone, this is a default setting, but you can also enable it manually by going into the Settings app > General > Accessibility > Toggle on "Tap to Wake."

- Make your device power efficient, because graphics tend to consume a lot of battery. Enable low power mode to save battery. Convert your phone to power saving mode by going into settings and use your phone longer. If your battery gets lower than 20%, it is automatically enabled, but if you want to enable it prior to that go to settings>>battery>>enable low power mode.

- If you have kids around and you want to make sure that your data is safe and at the same time friendly for the kids to use, then simply enable the friendly option. Simply go to Settings > General > Restrictions, and you can limit access to specific apps, block in-app purchases, and set an age range for appropriate content.

- Customize auto-correct text replacement. iOS 12 is quick at guessing what you might want to type next. In order to customize your kind of language you might want to use autocorrect. It can be an abbreviation to a word. Just go to Settings > General, scroll down and tap Keyboard. Select Text Replacement.

- Never fill in a password, address, or account information again with the iPhone XS. You can save time in filling the same account information again and again. You can enable your device to automatically detect the details you have already put in once, so when it is time to input the data, your iPhone auto fills the required information. Just Go to Settings > Safari > Auto fill

Phone Customization

- Use your device to its full potential even whilst you are multitasking or driving. Select the enable speak feature. With this option you can have your iPhone XS read out your texts. Go to Settings > General > Accessibility and toggle the option 'Speak Selection.' Long-press on the speech bubble within your Messages, and you'll now find the option to 'Speak.'

- Set up widgets which are frequently required by you. Widgets can be accessed if you swipe right from your notification center. Set up widgets by going to settings, click control center and see which widgets you want.

- You can set up widgets on the home screen or on the lock screen of your device. You can select weather, time, date, etc.

Chapter Eight
Multimedia

The iPhone XS can now adjust the photo depth in a picture thanks to the depth slider. The iPhone XS supports a vertical configuration camera with improved tone LED flash, accompanied with advanced flicker system in a 12 mega pixel camera with telephoto lens. The front camera has a two times faster sensor with improved red eye reduction along with detailed segmentation

The camera module of the iPhone XS is slightly larger. So if you put an XS case (a cover that is made by Apple because It has perfectly measured dimensions for each and every phone) on a 10s it would fit perfectly, and if you place the same XS case on iPhone 10, you will actually end up with a little gap around the bottom, so the extra room means that the camera module is a little smaller than the 10s.

Dual cameras use the difference in distance between the cameras to create a depth map. This segments the pictures into different layers. So, when you want to blur the background of a photo, instead of drawing against the foreground and then blurring the whole background, this portrait mode will blur the parts that are more distant and nearer parts to the camera less, which gives it a more realistic look. It's a subtle difference but a massive improvement. It gives a direction to the blur as well, such as a radial blur around the center of the frame.

Combined with the fact that it has the best video recording with 4k60 plus, Apple has added stereo audio recording with video, making a pretty powerful camera. New specs, new camera, this iPhone is everything that we thought it could be.

It also allows you to capture the full potential of your video recording on your iPhone. If you open the camera app and switch over to video, all your recording in is 1080p HD, but the rear camera can actually record 4k videos at 60 frames per second. 4k video looks amazing and is probably something you really want to try out. So, here's how you access this feature.

- Go to settings and scroll down to camera, and by default it will have selected 1080p at 30 frames per second. Select "4k at 60 frames per second." Now when you open the video camera in your iPhone you can record in amazing 4k

Camera

- Dual 12MP wide-angle and telephoto cameras
- Wide-angle: $f/1.8$ aperture
- Telephoto: $f/2.4$ aperture
- 2x optical zoom; digital zoom up to 10x

- Portrait mode with advanced bokeh and Depth Control
- Portrait Lighting with five effects (Natural, Studio, Contour, Stage, Stage Mono)
- Dual optical image stabilization
- Six-element lens
- Quad-LED True Tone flash with Slow Sync
- Panorama (up to 63MP)
- Sapphire crystal lens cover
- Backside illumination sensor
- Hybrid IR filter
- Autofocus with Focus Pixels
- Next-generation Neural Engine
- Tap to focus with Focus Pixels
- Smart HDR for photos
- Wide color capture for photos and Live Photos
- Local tone mapping
- Advanced red-eye correction
- Exposure control
- Auto image stabilization

- Burst mode
- Timer mode
- Photo geotagging
- Image formats captured: HEIF and JPEG

Display

- Super retina display (HD)
- inch (when measured diagonally) all screen OLED
- Multi touch
- HDR display
- 2688-by-1242-pixel resolution at 458 ppi
- 1,000,000:1 contrast ratio (typical)
- True Tone display
- Wide color display (P3)
- 3D Touch
- 625 cd/m2 max brightness (typical)
- Fingerprint-resistant oleophobic coating
- Support for display of multiple languages and characters simultaneously

Video

- 4K video recording at 24 fps, 30 fps, or 60 fps
- 1080p HD video recording at 30 fps or 60 fps
- 720p HD video recording at 30 fps
- Extended dynamic range for video up to 30 fps
- Optical image stabilization for video
- 2x optical zoom; digital zoom up to 6x
- Quad-LED True Tone flash
- Slo-mo video support for 1080p at 120 fps or 240 fps
- Time-lapse video with stabilization
- Cinematic video stabilization (1080p and 720p)
- Continuous autofocus video
- Take 8MP still photos while recording 4K video
- Playback zoom
- Video geotagging
- Video formats recorded: HEVC and H.264
- Stereo recording

True Depth Camera

- 7MP camera
- $f/2.2$ aperture
- Portrait mode with advanced bokeh and Depth Control
- Portrait Lighting with five effects (Natural, Studio, Contour, Stage, Stage Mono)
- Animoji and Memoji
- 1080p HD video recording at 30 fps or 60 fps
- Smart HDR for photos
- Extended dynamic range for video at 30 fps
- Cinematic video stabilization (1080p and 720p)
- Wide color capture for photos and Live Photos
- Retina Flash
- Backside illumination sensor
- Auto image stabilization
- Burst mode
- Exposure control
- Timer mode

Audio Playback

- Audio formats supported: AAC-LC, HE-AAC, HE-AAC v2, Protected AAC, MP3, Linear PCM, Apple Lossless, FLAC, Dolby Digital (AC-3), Dolby Digital Plus (E-AC-3), and Audible (formats 2, 3, 4, Audible Enhanced Audio, AAX, and AAX+)

- Wider stereo playback

- User-configurable maximum volume limit

Video Playback

- Video formats supported: HEVC, H.264, MPEG-4 Part 2, and Motion JPEG

- High Dynamic Range with Dolby Vision and HDR10 content

- AirPlay Mirroring, photos, and video out to Apple TV (2nd generation or later)

- Video mirroring and video out support: Up to 1080p through Lightning Digital AV Adapter and Lightning to VGA Adapter (adapters sold separately)

Chapter Nine
Using the Internet

In order to use applications that require internet, there are two options that you can use:

1. Over the Wi-Fi

    - Go to settings>>Wi-Fi>>Select the network you want to connect to >> enter the password >>then tap join

2. Using the cellular data

    - Go to settings>>Cellular>>Cellular data switch >>App >> Toggle the slider

Note: you can use cellular data for selective applications. Scroll to "use cellular data for" section to select only the desired applications.

Wi-Fi Hotspot

By using this feature, you will be able to use your internet with other devices as well. To set it up you need a sim card which has data option enabled.

Go to settings >> personal hotspot (this may not show up if your sim card which is basically set up by your carrier doesn't allow you to do Wi-Fi hotspot sharing>>toggle the personal hotspot to enable it. It has a default password which is complicated and random. You can change the password to make it easier for you to remember. While making the hotspot Wi-Fi visible to other devices it uses the phone's name which you can also change.

To change the name, go to settings>>general>>about>> name

Note: With iOS the hotspot times out after a couple of minutes so if you have something continuously connected to it, then it should continue to work.

Safari

Safari is a browser that is designed by Apple. It is a fast and efficient browser on which you can easily surf and shop online. You can prevent cross-site tracking; it basically prevents sites like advertisers from reading the cookies from websites you visit. In the end you will be seeing less targeted ads based on your browsing history

Have you ever needed to open multiple links in tabs from a single website and have to toggle between the tabs? In order to avoid this hassle, you have the feature in Safari where you can set links to open in the background, instead of popping up in a new tab, by heading to Safari settings and switching it to opening in the background.

You can use auto fill in Safari too. It's been around a while, but it gets even better with Face ID. It will save your information across multiple devices including brand new devices, so you don't have to enter your credentials all the time. To use it, first turn on key chain within iCloud settings then tap auto fill in Safari settings. Then put in your info and enable names and passwords. You can also save credit card information

Safari can also now automatically block annoying auto play videos. Just open Safari preferences, head to the website tab, then click auto play. You'll see an option for auto block settings when visiting other websites, which you can set to stop media with sound or never auto play. If you want to change the settings for a certain website just right click the address bar, click settings for this website, and then change the auto play setting.

If you always use the search engine to convert certain measurements, you can use Siri search instead. Just swipe down on the home screen and type in your conversion. The answer will instantly show up without you having to search. You can also look up a flight number to instantly track your flight.

Safari's reader mode has been around for a while and it's great because it blocks out ads. It only shows what you want to see. With Safari now you can automatically enable the reader mode when it's available. To turn it on just tap and hold the reader mode icon, then you can choose whether to use it only on that website or on all websites

You can also tap the font buttons to see the size and background color and font. With your iPhone XS, you can change the background color to black which will put less stress on your eyes and at the same time will conserve your battery too.

You can tap on and hold the bookmarks icon to either add a bookmark or add it to your reading list. To access your reading list just hit the bookmarks icon and then the reading list tab. There is also a history tab through which you can search your history. You can instantly open a new tab or a private tab by holding the tabs icon. Moreover, if you press the share button while viewing a page you have the option to print the page, find text within the page or create a PDF file of the webpage. After creating a PDF file, you can mark it up, share it, and save it to your device

When you create a new tab, you will see your favorites pop up along with your frequently visited sites. If you don't want one of your favorites tab to show up just tap and hold and delete it. If you don't want any frequently visited sites to show at all, you can turn that feature off in Safari settings. You can also clear all browsing history within the settings.

Mail

Mail Attachment Support

- Viewable document types
- .jpg, .tiff, .gif (images); .doc and .docx (Microsoft Word); .htm and .html (web pages); .key (Keynote); .numbers (Numbers); .pages (Pages); .pdf (Preview and

Adobe Acrobat); .ppt and .pptx (Microsoft PowerPoint); .txt (text); .rtf (rich text format); .vcf (contact information); .xls and .xlsx (Microsoft Excel); .zip; .ics; .usdz (USDZ Universal)

Open the settings app until you see accounts and select add account. You will be presented with seven account setup options:

- iCloud
- Exchange
- Google
- Yahoo
- Aol.
- Outlook.com
- Other

Once you have selected your required account, enter the correct email address and password then your account will be ready to use. You are going to access it in the mail app. By default, it should appear on your home screen. Based on the type of account, you will see different folders that you have the ability to explore such as inbox, drafts, sent, junk, trash, and any folders specific to the account.

To View Mail Settings

Go to settings>>Mail

Scroll down to the bottom of the page and select default account. This is important if you have more than one mail account, because when selecting new mail in the mail app, this is the default account that will automatically be chosen to send the email from.

Choosing how your mail is delivered to you from the settings page: scroll down to accounts and passwords and select fetch new data. You would want to change the settings to make sure that you receive your email in a way that works best for you. This page will allow you to control how the mail is downloaded to your device.

You will be presented with three options:

1. Push
2. Account list
3. Fetch

- Push mail: when turned on push automatically downloads from your account to your device as soon as you receive it in the server.

The alternative is that emails are only downloaded to your device when you check your mail app.

- Account list: Let's you choose to either fetch email automatically or only download mail when you check it. Tap each account and then tap fetch or the manual option.

Fetch is a more traditional way of checking email. It checks your email every 15, 30 or 60 minutes. It downloads the email that has arrived. You can also set it to check manually; this can be used if the push feature is disabled. The less often you check your email the more battery you will save.

There are certain setting in the mail application that let you control options, like:

1. Preview: the number of lines to show when the mail notification pops

2. Show to/CC: to show who emailed and who is CC'd in the email

3. Swipe options: control what happens when you swipe left or right across the email in the inbox field

4. Flag style: choose the type and color of flag for emails that you flagged

5. Ask before deleting: get a warning before an email is deleted

6. Load remote images: to load images in the email to save data

7. Organize by thread: to group together related messages from the same conversation

Chapter Ten
Phone Settings

Turn on Battery Percentage

If you see the battery bar at the top right of your screen, and every time you look at it makes you wonder how much battery is left, well you can get an accurate percentage for your battery now.

Enable the battery percentage by going to settings>> battery>>toggle the battery percentage on. However, to see the battery percentage, you have to swipe down to view the control center on the iPhone XS because of the notch, as it does not leave enough space to display the battery percentage on the screen.

How to Reset to Factory Settings

Experience lagging in your device? Changed the settings of your device and want to restore the default settings. Well you can easily restore the default settings by resetting the device.

Settings>> general>> rest >>erase all content from settings>>enter passcode>>erase iPhone.

Now if you have signed in to iCloud on your device, you will need to put in your iCloud password to remove iCloud before the reset. You can now set up your iPhone XS again from scratch.

How to Set Screen Lock Time

It is the amount of time before the screen completely shuts off.

Settings>>display & brightness>>Auto lock time.

You can manually set the amount of time that passes before shutdown. Keep in mind that this is a security function, so if you set this time too long there is more chance for someone to perhaps steal your phone or pick it up when you are not looking, or send an email or even access your personal information, so you need to set the auto lock time as per your convenience.

The longest time under the options would be "never," so unless you manually lock it every time, anyone who picks up your phone can access everything in your phone that is not password protected.

Hide Unwanted Apps

So, if you don't want anybody playing games on your phone or accessing private data, you can do the following:

Settings>>general>>restrictions (if you haven't used this section before on your iPhone XS you will need to enable this by creating a restriction passcode).

There are going to be a lot of different options on the screen, but scroll down to find the option that says "apps." In this you can turn off apps based on their age limit. It is recommended to select the option "Don't show apps" as it will hide everything. As soon as you do that and head over to the home screen, you will see that the apps are completely hidden.

The cool thing about this is that it will not delete the app, and it's not going to modify those apps. If you want them to show up again, go back to your restriction and enter the passcode that you earlier set up. Scroll back down to the "Apps" option, click allow all apps, and then it will place them back on the home screen.

Change Flashlight Intensity

Not every situation calls for a powerful beam. Bring up the control center and do a firm 3D touch on the flashlight icon. Now make the flash as bright or as dim as you like.

Auto Delete Unused Apps

Empower your phone to get rid of the apps you don't use. The setting is easy to turn on and if you regret not using an app, you can always re-install it.

Settings>>iPhone storage

Emoji Keyboard Replacement

Instead of searching through rows and rows of emoji to find the perfect one, try searching for it. Just hit the emoji button on your keyboard once you have typed a message. Tap any of the orange highlighted words to turn it into a pictogram.

Prioritize Certain Downloads

Tell your iPhone XS which downloads it should complete first. As your apps are updating apply a firm 3D touch press to your favorite and click prioritize download.

How to Set an iPhone Email Signature

Settings>>mail, contacts, calendar>>signature button.

Tap inside of the textbox to bring up keyboard. Edit the signature.

Change the Wallpaper on Your Device

Apple has default wallpapers that you can use for both home and lock screen. There are several wallpapers available, like dynamic wallpapers, which can move. There are other categories as well like still wallpapers and live wallpapers, or you can choose photos from the gallery as well. With live wallpapers, if you 3D touch them, they will start to move, which is pretty awesome and unique.

Access it by going to: Settings>>wallpapers.

Organizing Apps

You can organize your apps on the home screen. Click on an application and it will put all the applications to the wiggle mode and move them around. Just press and drag and you can move an app to a different location within the home screen.

Creating a Folder

If you want to put certain apps in a folder, you can easily do it by holding the specific app, combining it with the other, and your iPhone will automatically create a folder.

Chapter Eleven
Apps

Apple has inbuilt applications like Mail, Music, Wallet, Safari, maps, calendar, iTunes store, App Store, Notes, Contacts, Books, Home, Weather, Reminders, Clocks, Stocks, Calculator, Voice memo, Compass, Podcasts, Watch, Tips, Find my phone, Find my friends, Settings, File, Measure ,Pages, Numbers, Keynote, iMovie, GarageBand, iTunes, and Clips. These are all preinstalled.

- **Apple Pay**: In order to pay instantly, go to the Apple Pay application, tap the buttons on the right-hand side twice. This will instantly open the app for you and makes digital payments with easy access.

- **Quick-delete in the Calculator app**: A really good feature that has enabled using the calculator app in the iPhone effectively. Fortunately, now you can swipe across the number in the black area at the top - left or right, it doesn't matter - and for each swipe, a single digit will be removed from the end of the figure.

- **Stop music with a timer**: Doze off to your favorite playlist or podcast without letting it play all night. You might want to go to sleep while listening to music. Your phone will be playing songs throughout the night, and as a result it would consume your battery all night. You can easily turn on the timer and set what time you would like to stop the music. Open the Clock app's Timer tab. Select how long you want your timer to last for, and then press 'When Timer Ends.' Scroll down to the bottom of the menu and select 'Stop Playing.' Press start on the timer and then begin playing your music from the Music app. When the timer ends, the music will fade to a stop. This trick will also work for audiobooks and other media.

- **Apple Books**: iBook app is now Apple Books, and is redesigned with a more modern UI. The Apple bookstore app is really handy if you're searching for books that you like. Apple Books makes it very interesting to search for books within the app. It uses information such as authors, celebrities, or genres and makes it easy to look for your favorite books. You easily shuffle between the books you are reading on the app. You can also add books to your wish list in order to keep track of the books you want to read next.

- **Apple Music**: It is a fun and efficient way of listening to the music you like. The iPhone XS provides a music app which you can personalize according to the music genre you prefer listening to. You can use the lyrics of songs or an artist's name to find new music. This also recommends top songs on the daily charts based on your genre. Music app allows you to find a playlist that matches your mood. In for a workout and want to boost yourself, access the music app and find the workout playlist to keep moving to the rhythm.

- **Apple News**: for anyone who likes to keep an eye on the market frequently, the Apple News app will keep you up to date all the time. iOS 12 brings Apple News app that keeps track of your interests and suggests news related to it. It has become easier to look for content according to your likes and dislikes.

- **Voice Memos**: The iPhone XS brings you an all-new design with improved ease of use of memos by introducing audio memos. Why write when you can talk and make your notes, checklists, etc. iCloud keeps your recordings and edits in sync across all your devices and provides support on iPad for both portrait and landscape orientation.

- **Television**: The app notifies you whenever a new episode of your favorite TV shows are uploaded. You can also share them universally. It shows suggestions based on your watch history and genres you watch.

- **Apple Podcasts**: This feature now supports chapters for shows that include them. You can skip 30 seconds or to the next chapter with forward and back buttons in your car or on your headphones. This leads the user to easily manage new episode notifications from the Listen Now screen.

- **Accessibility**: Live Listen now works with Air Pods to help you hear more clearly. RTT phone calling now works with AT&T. Speak Selection now supports using the Siri voice to speak text that you have selected when you don't have time to read them yourself.

- **Photos**: This feature helps the user in rediscovering and sharing the photos that are synced with your library.

- **Screen Time**: This feature assists the user by informing them about how they spend their time on their phone. The user can analyze their screen time better and cut down accordingly. Directly addressing how you use your phone and how much you use your phone gives an analysis on whether you are addicted to your phone or not. It also tells you what application has been used the most from the past few hours or throughout the day. It also lets you set limits on apps. Let's say you want to spend only one hour on Instagram, once that hour is up it doesn't let you in unless you click on the button that lets you in. It gives a detailed breakdown of the usage of the apps. How many times you pick up your phone during the day, the average per hour and which apps send you the most notifications over the course of the day

- **Siri**: In addition of Siri, the user can easily create customizable shortcuts for carrying out tasks on their devices. Siri shortcuts are also available. Essentially it will allow you to map any in app action/command, to any phrase you want: e.g. check the Apple stock price. I chose the, "Show me money." When I say this phrase to Siri it opens the stock app

- **Improved notification manager**: iOS 12 has improved notification tools that only notify you about important tasks, as per your command, instead of spam notifying you with unnecessary alerts.

- **Camera tool**: The hyped change in the iPhone XS Max is the addition of FaceTime, however this feature has been removed from the iOS 12 update for the time being.

- **App management**: With the impressive iOS 12, now there is no need to close apps manually; Apple brings the facility to automatically close apps that are not in use and save battery.

- **Built-in measuring tool**: Now you can measure objects by placing it in front of your camera. The new app measure is based on augmented reality that literally lets you look through the camera and measure that object e.g. how long or how tall the object is or how much volume it has. The only limitation is that it is better with small objects and things that are closer, to you but it is definitely accurate**.**

- **Bedtime**: Get those irregular sleeping cycle's right. The iPhone helps to build a healthy bedtime routine.

When bedtime starts the light automatically dims and notifications are hidden until you unlock the phone in the morning. To access bedtime, go to clock >> bedtime and follow the instructions to get started.

Chapter Twelve
Tips, Tricks, and Shortcuts

Best hidden features/tips and tricks for you to use your iPhone to its highest potential

- Access dark mode: where all blacks turn to white and all whites turn to black. Now why is this useful? Well, let's say you are reading an article at night and wanted to go a little bit easier on the eyes. Now, the iPhone has the feature to switch to night mode and the best part is it only inverts the whites to blacks and vice versa.

  This mode can be achieved just by triple clicking the side button. First you need to set it up. Hop into settings and scroll down to general, then click on accessibility, and once you're in and scroll down until you see accessibility shortcut. From the options provided, click on the "smart inverted colors." Make sure you do this instead of "classic invert colors" which inverts all the colors.

  If you use Safari or are reading an iBook in dark mode, it will look exactly the same as white to black and vice versa.

- Add a virtual home button: on iPhone 10s. This is really cool as they don't have a home button. To access an app or other features like Siri, you have to use one of the following:

- 3D touch: a slightly stronger touch will exit from the app just like the regular home button. If you touch it slightly it pops up with a lot of different options e.g., view your notifications, rotate the screen, increase and decrease the volume, etc.

- To adjust the sensitivity of the 3D touch, go to settings, click on accessibility, then click 3D touch and choose light medium or firm touch.

- There is a very easy way to set this up on your device. What you are looking for is assistive touch. Turn it on and it pops on with the virtual home button. You can choose gestures, and there are customization options available to look at. E.g. single tap to open menu, double tap for Siri. If you long press it, it acts as a regular home button.

- Use your device single handedly, which is very useful if you want to type with one hand, especially when you are on the go. It's very easy to set up. All you need to do is click the globe button by 3D touch and you are able to swipe up and choose the one-handed keyboard.

- The next feature is very serious as it can save you from a dangerous situation. This is called emergency SOS. What it lets you do is press the side button five time in order to call the emergency services discretely and safely in order to get you help if you need it. I encourage anyone reading this manual just to have it in case.

Setup: settings>>emergency SOS >> turn on call with side button by default. Apple does have the emergency feature if you press the side button and either of the volume buttons.

- Record video of your iPhone screen: Access the screen recording button in the control center. Once you press it, it will begin a countdown and then it will start recording the screen of your iPhone XS. In the top left you will be able to see that there is a red box, with recording time as an indicator of your screen being recorded. This is best if you want to record a game. Tap the time button once you have completed recording the screen and click stop recording and it will then automatically save that to your photos.

- Turn your Apple ear pods into live listening hearing aids that cancel out background noise and work incredibly well if you are in a crowded space. Enable the live listening hearing. Once you have your ear pods connected to your iPhone, press the live listen option and you will be able to turn your device into a hearing aid. You can magnify the volume, so if you are always on the go this feature will work wonders.

- Keep a track of all your healthy and unhealthy habits with the help of the in-built app that not only records your activity during the day but also counts the number of hours you sleep. Count your steps: iPhone counts the number of steps you've taken or for how long you've run. If you want to see how you are doing open the health app and you will be able to see your daily, weekly, monthly, and yearly reports.

- To access the features of the control center swipe down from the top right. You can view the basic functions and add more features in the control center. Go to settings, click on control center, now click customize additional controls that appear in the control center. You can add another great feature that can be used whilst driving. It detects when you are driving and automatically sets to do not disturb mode. Another control feature that you can add is screen recording, etc.

- Text size: you can change the size of the text on your phone. Change the size of the text really quickly by putting it in the control center.

- The app switcher is another great feature to maximize the use of your device. The physical home button on the latest iPhone XS has been replaced by a home button on the multi-dimensional touch screen. In order to switch between apps effectively and easily, you can swipe that home button towards the right and it will take you directly to the app switcher, from where you can easily switch between apps.

- Zoom into the YouTube App: In order to take full advantage of large HD displays, Apple has added a feature where you can zoom into YouTube videos, and spread them over the landscape display canvas by merely pinching into the screen.

- Do not disturb: This is a feature that lets you keep your personal time to yourself without being interrupted by phone calls or messages. You can easily customize the time period you don't want to be disturbed between. After you set the time, the iPhone will automatically go into "Do Not Disturb" mode during that specific time.

- Close apps that are using more space and battery to maximize the battery life. Do this every now and then.

- Tap to top: this is a treat for people who read a lot of documents and books on their device. While going through long notes and documents, going back to the top seems like a timely hassle. To save yourself time, just tap the very top of your iPhone XS Max screen and it will take you back to the very first page of the document.

- Quickly add symbols: In order to use the iPhone XS Max keyboard more effectively and make typing quicker and easier, instead of tapping once on the number button, once on your chosen symbol, and then once again on the alphabet button to go back to the conventional keyboard layout, you can do the whole thing in one gesture. Just press the number button and, without taking your finger off, slide to the symbol you want to add and release. The keyboard will know to go back to the original form itself. One symbol if long pressed will also offer other symbols that you might find are absent on the keyboard. They are all there, hidden in the keyboard.

- Rich formatting: rich formatting is a trick to make certain parts of your app texts stand out and be different. Just open an app that supports rich formatting, highlight the text you'd like to edit by double-tapping it, and select the formatting menu.

- In order to maintain your device, it is highly recommended you power off your iPhone XS at least once a month. And that means completely shutting it off. The way to do this is press the side the button and the volume up button at the same time. It will bring up the menu and offer two options: "slide to power off" and "emergency SOS."

- In the XS model, a smart HDRA is powered by the same A12 chip that allows the phone's image signal processor and neural engine to chain multiple pictures into one by using techniques like zero shutter lag, resulting in better picture quality. The camera is now also able to take improved photos and videos with greater highlight structure in low-light conditions.

Chapter Thirteen
Conclusion

The iPhone XS is Apple's flagship smartphone and is a slight upgrade from the iPhone X. Externally, it is identical to last year's model and has the same screen size. However, the new iPhone XS has an improved screen resolution, better camera technology, and a more advanced A12 chipset. In fact, the new processor is 15% faster and consumes 40% less energy.

The phone's second iteration, the iPhone XS Max has the largest screen for any Apple smartphone and can be categorized as a 'phablet' to compete with Samsung's popular Note series.

The iPhone XS still has plenty of novelty features such as waterproofing, wireless charging and 3D touch, all of which make it more fun to use.

Apple also claims to have improved the security features of the phone including a faster face ID. This not only provides more privacy but also makes it user-friendly. There are also several 'hidden' features which enhance the phone's usability among new Apple users.

For people who did not buy last year's iPhone X, the new model is a fantastic upgrade. All in all, this is an expensive phone but easily one of the most advanced on the market.

www.ingramcontent.com/pod-product-compliance
Lightning Source LLC
Chambersburg PA
CBHW071417220526
45469CB00004B/1318